PRESSURE VESSEL INTERVIEW QUESTIONS AND ANSWERS

FOR A SUCCESSFUL CAREER IN PRESSURE VESSEL TECHNOLOGY

CHETAN SINGH

Made with ♥ on the Notion Press Platform
www.notionpress.com

Contents

Acknowledgements

Writing this book on Pressure Vessel Interview Questions and Answers was a journey filled with the support and guidance of many individuals. First and foremost, I would like to express my sincere gratitude to all the professionals in the field of pressure vessels who shared their knowledge and experience in answering the questions included in this book. Your contributions have been invaluable and have made this book a comprehensive resource for anyone looking to excel in their career in this field.

I would also like to thank my family for their unwavering support and encouragement during the writing process. Their love and support have been a source of inspiration and have helped me to overcome the challenges that came with writing this book.

This book is a testament to the hard work and dedication of many individuals, and I am honored to have had the opportunity to play a part in its creation.

CHAPTER ONE

Introduction:

Pressure vessels are critical components of various industrial processes and are used to store and transport various liquids, gases, and chemicals under pressure. As a result, pressure vessels play a crucial role in ensuring the safety and efficiency of these processes.

For individuals seeking a career in this field, it is essential to have a good understanding of the design, fabrication, inspection, and maintenance of pressure vessels. In order to assist with this, this book provides a comprehensive collection of interview questions and answers related to pressure vessels.

The questions and answers in this book have been compiled by experienced professionals in the field and are designed to test and enhance your understanding of pressure vessels. Whether you are a student, an engineer, or a seasoned professional, this book is an excellent resource to prepare you for any interview in the field of pressure vessels.

By studying the questions and answers in this book, you will not only become more knowledgeable about pressure vessels, but you will also become better prepared to tackle interview questions and demonstrate your expertise in this field. So if you want to excel in your career in pressure vessels, grab a copy of this book today and start preparing!

CHAPTER TWO

Basic interview based Pressure Vessel Questions and Answers

What is a pressure vessel?

A pressure vessel is a container designed to hold gases or liquids at a pressure substantially different from the ambient pressure.

What are some common materials used to make pressure vessels?

Some common materials used to make pressure vessels include carbon steel, stainless steel, aluminum, and fiberglass reinforced plastic (FRP). The material selection depends on the type of fluid being contained, the design pressure and temperature, and the corrosion resistance requirements.

What are some common applications of pressure vessels?

Pressure vessels are used in a wide range of applications, including the storage of gases (such as air, oxygen, and natural gas) and liquids (such as water, oil, and chemicals). They are also used in the chemical and petrochemical industry, power plants, and the food and beverage industry.

What are the design codes and standards that apply to pressure vessels?

The design and construction of pressure vessels are governed by various codes and standards, such as the ASME Boiler and Pressure

Vessel Code (ASME BPVC) in the United States, the PED (Pressure Equipment Directive) in Europe, and the Australian Pressure Vessels Code (AS 4041). These codes provide guidelines for the design, fabrication, inspection, and testing of pressure vessels to ensure their safety and reliability.

How do you inspect and test a pressure vessel?

Pressure vessels are inspected and tested at various stages of their life cycle, including during fabrication, after repair or modification, and at regular intervals during operation. Some common inspection and testing methods include visual inspection, radiographic inspection, ultrasonic inspection, hydrostatic testing, and pneumatic testing. These tests are carried out to verify that the vessel is free of defects, meets the design specifications, and is safe to use.

What are the different types of pressure vessels?

Pressure vessels can be classified based on their shape, size, and intended use. Some common types of pressure vessels include:

- Spherical pressure vessels: These have a spherical shape and are used for storing liquids or gases at high pressure.
- Cylindrical pressure vessels: These have a cylindrical shape and are used for storing liquids or gases at moderate pressure.
- Compressed air tanks: These are used to store compressed air at high pressure.
- Gas cylinders: These are used to store gases at high pressure, such as oxygen, hydrogen, and nitrogen.
- Boilers: These are used to generate steam by heating water.
- Autoclaves: These are used to sterilize equipment and materials by exposing them to high pressure steam.

What are some factors that can affect the strength and reliability of a pressure vessel?

Some factors that can affect the strength and reliability of a pressure vessel include the material selection, design pressure and temperature, corrosion, fatigue, and weld quality. It is important to consider these factors during the design and construction of a pressure vessel to ensure its safe and reliable operation.

What are some common problems that can occur with pressure vessels?

Some common problems that can occur with pressure vessels include corrosion, fatigue, and overpressure. Corrosion can weaken the vessel and reduce its service life, while fatigue can cause cracks to form due to the repeated loading and unloading of the vessel. Overpressure can occur if the vessel is subjected to a pressure greater than its design pressure, which can lead to failure.

What are some safety measures that should be taken when working with pressure vessels?

Some safety measures that should be taken when working with pressure vessels include ensuring that the vessel is properly designed, constructed, and maintained, following proper operating procedures, and using protective equipment (such as safety glasses, gloves, and hard hats). It is also important to have emergency procedures in place in case of an accident or failure.

How do you calculate the minimum required thickness of a pressure vessel wall?

The minimum required thickness of a pressure vessel wall can be calculated using the following formula:

$t = PD/SE$

where t is the minimum required thickness, P is the maximum allowable design pressure, D is the vessel diameter, and SE is the allowable stress at the design temperature.

How do you calculate the maximum allowable pressure for a pressure vessel?

The maximum allowable pressure for a pressure vessel can be calculated using the following formula:

$P = SE*t/D$

where P is the maximum allowable pressure, SE is the allowable stress at the design temperature, t is the vessel wall thickness, and D is the vessel diameter.

How do you calculate the volume of a pressure vessel?

The volume of a pressure vessel can be calculated using the following formula:

$V = \pi r^2 L$

where V is the volume, r is the radius of the vessel, and L is the length of the vessel.

How do you calculate the weight of a pressure vessel?

The weight of a pressure vessel can be calculated using the following formula:

$W = \rho V g$

where W is the weight, ρ is the density of the material, V is the volume, and g is the acceleration due to gravity.

How do you calculate the diameter of a pressure vessel?

The diameter of a pressure vessel can be calculated using the following formula:

$D = 2 * r$

where D is the diameter and r is the radius of the vessel.

What are the different types of heads used on pressure vessels?

Pressure vessels can have different types of heads, depending on the shape and size of the vessel and the intended use. Some common types of heads used on pressure vessels include:

Hemispherical heads: These have a half-spherical shape and are used for high pressure vessels.

Torispherical heads: These have a shape that is a combination of a sphere and a torus (doughnut) and are used for moderate pressure vessels.

Ellipsoidal heads: These have an ellipsoidal shape and are used for moderate pressure vessels.

Conical heads: These have a conical shape and are used for low pressure vessels.

Flat heads: These have a flat shape and are used for low pressure vessels.

What are the different types of closures used on pressure vessels?

Pressure vessels can have different types of closures, depending on the intended use and the pressure and temperature conditions.

Some common types of closures used on pressure vessels include:

- Flanged closures: These are used to seal the vessel and are held in place by bolts.
- Threaded closures: These are used to seal the vessel and are held in place by threads.
- Welded closures: These are used to seal the vessel and are held in place by welding.
- Clamped closures: These are used to seal the vessel and are held in place by clamps.

What are the different types of supports used for pressure vessels?

Pressure vessels can be supported in a number of ways, depending on the size and shape of the vessel and the intended use. Some common types of supports used for pressure vessels include:

Skirt supports: These are used to support the vessel at the bottom and provide stability.

Leg supports: These are used to support the vessel at the bottom and provide stability.

Saddle supports: These are used to support the vessel at the bottom and provide stability.

Ring supports: These are used to support the vessel at the top and provide stability.

Pipe supports: These are used to support the vessel at the top and provide stability.

What are the different types of nozzles used on pressure vessels?

Pressure vessels can have different types of nozzles, depending on the intended use and the pressure and temperature conditions. Some common types of nozzles used on pressure vessels include:

- Inlet nozzles: These are used to introduce fluids into the vessel.
- Outlet nozzles: These are used to discharge fluids from the vessel.

- Manway nozzles: These are used to provide access to the inside of the vessel.
- Connection nozzles: These are used to connect the vessel to other equipment.
- Vent nozzles: These are used to release pressure from the vessel.

What are the different types of supports used for pressure vessels?

Pressure vessels can be supported in a number of ways, depending on the size and shape of the vessel and the intended use. Some common types of supports used for pressure vessels include:

- Skirt supports: These are used to support the vessel at the bottom and provide stability.
- Leg supports: These are used to support the vessel at the bottom and provide stability.
- Saddle supports: These are used to support the vessel at the bottom and provide stability.
- Ring supports: These are used to support the vessel at the top and provide stability.
- Pipe supports: These are used to support the vessel at the top and provide stability.

What are the different types of foundations used for pressure vessels?

Pressure vessels can be supported on different types of foundations, depending on the size and weight of the vessel and the intended use. Some common types of foundations used for pressure vessels include:

Concrete foundations: These are used to support heavy vessels and provide stability.

- Steel foundations: These are used to support heavy vessels and provide stability.

- Wood foundations: These are used to support light vessels and provide stability.
- Grout foundations: These are used to support vessels and provide stability.
- Isolation foundations: These are used to reduce the transmission of vibration and noise from the vessel to the surrounding structure.

What are the different types of coatings used on pressure vessels?

Pressure vessels can have different types of coatings applied to the surface to protect against corrosion, wear, and other types of damage. Some common types of coatings used on pressure vessels include:

- Epoxy coatings: These are used to provide excellent corrosion resistance and adhesion.
- Polyurethane coatings: These are used to provide excellent corrosion resistance and flexibility.
- Zinc coatings: These are used to provide excellent corrosion resistance and are often used as a base coat for other coatings.
- Rubber coatings: These are used to provide excellent abrasion resistance and are often used on nozzles and other high wear areas.
- Teflon coatings: These are used to provide excellent wear resistance and are often used on nozzles and other high wear areas.

What are the different types of gaskets used in pressure vessels?

Gaskets are used to seal the joints between the different parts of a pressure vessel and prevent the escape of fluids. Some common types of gaskets used in pressure vessels include:

- Spiral wound gaskets: These are made of a spiral wound strip of metal and a soft filler material and are used for high pressure and

temperature applications.
- Kammprofile gaskets: These are made of metal and have a wave-shaped profile, which allows them to seal against uneven surfaces.
- Ring joint gaskets: These are used to seal flanged joints and are made of metal with a specific shape that allows them to seal against the flange faces.
- Sheet gaskets: These are made of flat sheets of material and are used for low pressure and temperature applications.
- Metal jacketed gaskets: These are made of a metal core with a softer material jacket and are used for high pressure and temperature applications.

What are the different types of bolts used in pressure vessels?

Bolts are used to hold the different parts of a pressure vessel together and provide structural support. Some common types of bolts used in pressure vessels include:

- Hex bolts: These have a hexagonal head and are used to fasten parts together.
- Stud bolts: These have threads at both ends and are used to fasten parts together.
- Tap bolts: These have a hexagonal head and are used to fasten parts together in tapped holes.
- Machine bolts: These have a variety of head shapes and are used to fasten parts together.
- Structural bolts: These have a heavy hex head and are used to fasten structural members together.

What are the different types of flanges used in pressure vessels?

Flanges are used to connect pressure vessels to other equipment and provide a means of sealing the joints. Some common types of flanges used in pressure vessels include:

- Weld neck flanges: These have a long neck and are used to provide a smooth transition between the flange and the vessel.
- Slip-on flanges: These have a short neck and are used to provide a quick and easy connection.
- Threaded flanges: These have threads on the inside and are used to connect to threaded pipes.
- Socket weld flanges: These have a small bore and are used to connect to small diameter pipes.
- Blind flanges: These have no bore and are used to seal off the end of a pipe or vessel.

What are the different types of welds used in pressure vessels?

Welds are used to join the different parts of a pressure vessel and provide structural support. Some common types of welds used in pressure vessels include:

- Groove welds: These are used to join two parts that have a joint with a groove between them.
- Fillet welds: These are used to join two parts that have a joint with a gap between them.
- Plug welds: These are used to join two parts that have a hole in one of the parts.
- Slot welds: These are used to join two parts that have a slot in one of the parts.
- Flange welds: These are used to join two flanges together.

What are the different types of heat exchangers used in pressure vessels?

Heat exchangers are used to transfer heat between two fluids and are commonly used in pressure vessels. Some common types of heat exchangers used in pressure vessels include:

- Shell and tube heat exchangers: These have a series of tubes that contain the fluid being heated or cooled, and a shell that contains the heating or cooling fluid.

- Plate heat exchangers: These have a series of metal plates that are used to transfer heat between the two fluids.
- Spiral heat exchangers: These have a series of spiral channels that are used to transfer heat between the two fluids.
- Finned tube heat exchangers: These have tubes with fins attached to them, which increase the surface area and improve the heat transfer.
- Air cooled heat exchangers: These use air as the cooling fluid and are often used in outdoor applications.

What are the different types of pumps used in pressure vessels?

Pumps are used to move fluids through a pressure vessel and are commonly used in various applications. Some common types of pumps used in pressure vessels include:

- Centrifugal pumps: These use a spinning impeller to create a high-velocity fluid flow.
- Positive displacement pumps: These use a reciprocating or rotary motion to move the fluid.
- Diaphragm pumps: These use a flexible diaphragm to move the fluid.
- Piston pumps: These use a reciprocating motion to move the fluid.
- Screw pumps: These use a rotating screw to move the fluid.

What are the different types of valves used in pressure vessels?

Valves are used to control the flow of fluids in and out of a pressure vessel and are commonly used in various applications. Some common types of valves used in pressure vessels include:

- Gate valves: These have a sliding gate that opens and closes to control the flow of fluid.
- Globe valves: These have a movable disc that opens and closes to control the flow of fluid.

- Ball valves: These have a ball with a hole in the center that rotates to control the flow of fluid.
- Butterfly valves: These have a disc that rotates to control the flow of fluid.
- Check valves: These allow fluid to flow in one direction and prevent it from flowing in the opposite direction.

What are the different types of relief valves used in pressure vessels?

Relief valves are used to protect a pressure vessel from overpressure by releasing excess fluid or gas when the pressure exceeds a certain level. Some common types of relief valves used in pressure vessels include:

- Spring-loaded relief valves: These use a spring to open and close the valve based on the pressure.
- Weight-loaded relief valves: These use a weight to open and close the valve based on the pressure.
- Pilot-operated relief valves: These use a smaller valve to open and close the main valve based on the pressure.
- Rupture disks: These are thin metal disks that burst at a predetermined pressure, allowing the fluid or gas to escape.

What are the different types of insulation used in pressure vessels?

Insulation is used to reduce the heat transfer in pressure vessels and is commonly used in various applications. Some common types of insulation used in pressure vessels include:

- Fiberglass insulation: This is made of glass fibers and is used to insulate pipes and vessels.
- Mineral wool insulation: This is made of fibers of minerals, such as rock or slag, and is used to insulate pipes and vessels.
- Foam insulation: This is made of foam and is used to insulate pipes and vessels.

- Reflective insulation: This is made of reflective material and is used to reduce heat transfer by reflecting the heat back to the source.
- Cryogenic insulation: This is used to insulate pipes and vessels that are handling very low temperature fluids, such as liquid nitrogen or liquid helium.

How do you calculate the corrosion allowance for a pressure vessel?

Corrosion allowance is the additional thickness added to a pressure vessel to account for corrosion over its service life. The corrosion allowance can be calculated using the following formula:

Corrosion allowance = corrosion rate * expected service life * material thickness

where corrosion rate is the rate of corrosion in inches per year, expected service life is the expected lifespan of the vessel in years, and material thickness is the thickness of the vessel in inches.

How do you calculate the fatigue allowance for a pressure vessel?

Fatigue allowance is the additional thickness added to a pressure vessel to account for the effects of cyclic loading on the vessel's service life. The fatigue allowance can be calculated using the following formula:

Fatigue allowance = (Kf * Sf * Nf) / (Smax - Smin)

where Kf is the fatigue strength coefficient, Sf is the fatigue strength, Nf is the number of cycles to failure, Smax is the maximum stress, and Smin is the minimum stress.

How do you calculate the corrosion and fatigue allowances for a pressure vessel?

To calculate the corrosion and fatigue allowances for a pressure vessel, you can use the following formula:

Total allowance = corrosion allowance + fatigue allowance

where corrosion allowance is the additional thickness added to the vessel to account for corrosion over its service life, and fatigue allowance is the additional thickness added to the vessel to account for the effects of cyclic loading on the vessel's service life.

What are the different types of non-destructive testing methods used for pressure vessels?

Non-destructive testing (NDT) methods are used to evaluate the condition of a pressure vessel without causing any damage to the vessel. Some common types of NDT methods used for pressure vessels include:

- Radiographic testing: This involves using X-rays or gamma rays to create an image of the inside of the vessel and detect any defects or imperfections.
- Ultrasonic testing: This involves using high-frequency sound waves to inspect the vessel for defects or imperfections.
- Magnetic particle testing: This involves using a magnetic field to detect surface defects in ferromagnetic materials.
- Liquid penetrant testing: This involves applying a liquid to the surface of the vessel and using a developer to reveal any surface defects or imperfections.
- Eddy current testing: This involves using an electromagnetic field to detect defects or imperfections in conductive materials.

What are the different types of materials used for pressure vessels?

Pressure vessels can be made of various materials, depending on the intended use and the pressure and temperature conditions. Some common materials used for pressure vessels include:

- Carbon steel: This is a common material used for pressure vessels and is strong and durable.
- Stainless steel: This is a corrosion-resistant material that is commonly used for pressure vessels.
- Aluminum: This is a lightweight material that is commonly used for pressure vessels.
- Nickel alloys: These are corrosion-resistant materials that are commonly used for pressure vessels.

- Copper alloys: These are corrosion-resistant materials that are commonly used for pressure vessels.

What are the different types of welded joints used in pressure vessels?

Welded joints are used to connect the different parts of a pressure vessel and provide structural support. Some common types of welded joints used in pressure vessels include:

- Butt welds: These are welds that are used to join two parts together end to end.
- T-welds: These are welds that are used to join two parts together at a right angle.
- Lap welds: These are welds that are used to join two parts together by overlapping them.
- Corner welds: These are welds that are used to join two parts together at a right angle.
- Fillet welds: These are welds that are used to join two parts together with a gap between them.

What are the different types of non-welded joints used in pressure vessels?

Non-welded joints are used to connect the different parts of a pressure vessel and provide structural support. Some common types of non-welded joints used in pressure vessels include:

- Bolted joints: These are joints that are held together using bolts.
- Riveted joints: These are joints that are held together using rivets.
- Pinned joints: These are joints that are held together using pins.
- Screwed joints: These are joints that are held together using screws.
- Clamped joints: These are joints that are held together using clamps.

What are the factors that need to be considered when designing a pressure vessel?

There are several factors that need to be considered when designing a pressure vessel, including:

- The type and properties of the fluid or gas that will be contained in the vessel.
- The operating temperature and pressure of the vessel.
- The material of construction of the vessel.
- The size and shape of the vessel.
- The intended use of the vessel.
- The required strength and stability of the vessel.
- Any regulatory requirements or codes that need to be followed.

What are the different types of pressure vessel codes and standards?

There are various codes and standards that provide guidelines and requirements for the design, fabrication, inspection, and testing of pressure vessels. Some common codes and standards include:

- ASME Boiler and Pressure Vessel Code: This is a widely used code that provides requirements for the design, fabrication, inspection, and testing of boilers and pressure vessels.
- ISO 4126: This is an international standard that provides guidelines for the design, fabrication, inspection, and testing of pressure vessels.
- PED (Pressure Equipment Directive): This is a European Union directive that provides requirements for the design, fabrication, inspection, and testing of pressure vessels.
- Australian Pressure Vessel Code: This is an Australian standard that provides requirements for the design, fabrication, inspection, and testing of pressure vessels.
- JIS B 8241: This is a Japanese standard that provides requirements for the design, fabrication, inspection, and testing of pressure vessels.

What are the different types of pressure vessel materials?

Pressure vessels can be made of various materials, depending on the intended use and the pressure and temperature conditions. Some common materials used for pressure vessels include:

- Carbon steel: This is a common material used for pressure vessels and is strong and durable.
- Stainless steel: This is a corrosion-resistant material that is commonly used for pressure vessels.
- Aluminum: This is a lightweight material that is commonly used for pressure vessels.
- Nickel alloys: These are corrosion-resistant materials that are commonly used for pressure vessels.
- Copper alloys: These are corrosion-resistant materials that are commonly used for pressure vessels.

What are the different types of pressure vessel supports?

Pressure vessel supports are used to provide structural support to the vessel and prevent it from collapsing under the internal pressure. Some common types of pressure vessel supports include:

- Legs: These are supports that are attached to the bottom of the vessel and provide stability.
- Saddles: These are supports that are attached to the bottom of the vessel and provide stability.
- Skirts: These are supports that are attached to the bottom of the vessel and provide stability.
- Stands: These are supports that are used to elevate the vessel off the ground.
- Frames: These are supports that are used to provide structural support to the vessel.

What are the different types of pressure vessel closures?

Pressure vessel closures are used to seal the openings in the vessel and prevent the escape of fluid or gas. Some common types of pressure vessel closures include:

- Flanges: These are circular rings that are used to seal the openings in the vessel.
- Caps: These are covers that are used to seal the openings in the vessel.
- Plugs: These are inserts that are used to seal the openings in the vessel.
- Lids: These are covers that are used to seal the openings in the vessel.
- Covers: These are covers that are used to seal the openings in the vessel.

What are the different types of pressure vessel nozzles?

Pressure vessel nozzles are used to connect the vessel to other equipment or piping and provide a means of fluid or gas flow. Some common types of pressure vessel nozzles include:

- Flanged nozzles: These are nozzles that are connected to the vessel using a flange.
- Threaded nozzles: These are nozzles that are connected to the vessel using threads.
- Welded nozzles: These are nozzles that are welded to the vessel.
- Clamped nozzles: These are nozzles that are connected to the vessel using clamps.
- Socket weld nozzles: These are nozzles that are connected to the vessel using a socket weld.

What are the different types of pressure vessel linings?

Pressure vessel linings are used to protect the vessel from corrosion or other damage and are commonly used in various applications. Some common types of pressure vessel linings include:

- Glass linings: These are linings made of glass and are used to protect the vessel from corrosion.
- Rubber linings: These are linings made of rubber and are used to protect the vessel from corrosion and wear.
- Plastic linings: These are linings made of plastic and are used to protect the vessel from corrosion and wear.
- Ceramic linings: These are linings made of ceramic and are used to protect the vessel from corrosion and wear.
- Metal linings: These are linings made of metal and are used to protect the vessel from corrosion and wear.

What are the different types of pressure vessel coatings?

Pressure vessel coatings are used to protect the vessel from corrosion or other damage and are commonly used in various applications. Some common types of pressure vessel coatings include:

- Paint coatings: These are coatings made of paint and are used to protect the vessel from corrosion and wear.
- Powder coatings: These are coatings made of dry powder and are used to protect the vessel from corrosion and wear.
- Galvanizing: This is a coating process that involves applying a layer of zinc to the vessel to protect it from corrosion.
- Electroplating: This is a coating process that involves applying a layer of metal to the vessel using an electric current.
- Anodizing: This is a coating process that involves applying a layer of oxide to the vessel to protect it from corrosion.

What are the different types of pressure vessel testing?

Pressure vessel testing is used to evaluate the safety and performance of the vessel and ensure that it meets the required specifications. Some common types of pressure vessel testing include:

- Hydrostatic testing: This is a type of testing that involves filling the vessel with water and applying pressure to test the vessel's strength and integrity.
- Pneumatic testing: This is a type of testing that involves filling the vessel with air and applying pressure to test the vessel's strength and integrity.
- Burst testing: This is a type of testing that involves pressurizing the vessel until it fails to test its maximum pressure capacity.
- Leak testing: This is a type of testing that involves filling the vessel with a liquid or gas and checking for leaks to test the vessel's integrity.
- Ultrasonic testing: This is a type of testing that involves using high-frequency sound waves to inspect the vessel for defects or imperfections.

What are the different types of pressure vessel repair?

Pressure vessel repair is the process of repairing or replacing damaged or worn parts of the vessel to ensure its safety and performance. Some common types of pressure vessel repair include:

- Weld repair: This involves repairing damaged or worn parts of the vessel using welding.
- Cold repair: This involves repairing damaged or worn parts of the vessel using cold working techniques, such as bending or stretching.
- Hot repair: This involves repairing damaged or worn parts of the vessel using hot working techniques, such as forging or casting.
- Machining repair: This involves repairing damaged or worn parts of the vessel using machining techniques, such as turning or milling.
- Plating repair: This involves repairing damaged or worn parts of the vessel using plating techniques, such as electroplating or galvanizing.

What are the different types of pressure vessel maintenance?

Pressure vessel maintenance is the process of inspecting and maintaining the vessel to ensure its safety and performance. Some common types of pressure vessel maintenance include:

- Visual inspection: This involves visually inspecting the vessel for any signs of damage or wear.
- Non-destructive testing: This involves using testing methods, such as radiographic testing or ultrasonic testing, to inspect the vessel for defects or imperfections.
- Pressure testing: This involves testing the vessel's pressure capacity to ensure it is safe to use.
- Cleaning: This involves cleaning the vessel to remove any dirt or debris.
- Lubrication: This involves lubricating the moving parts of the vessel to reduce wear and tear.
- Repair and replacement: This involves repairing or replacing any damaged or worn parts of the vessel.

What are the different types of pressure vessel inspections?

Pressure vessel inspections are used to evaluate the safety and performance of the vessel and ensure that it meets the required specifications. Some common types of pressure vessel inspections include:

- Visual inspection: This involves visually inspecting the vessel for any signs of damage or wear.
- Non-destructive testing: This involves using testing methods, such as radiographic testing or ultrasonic testing, to inspect the vessel for defects or imperfections.
- Dimensional inspection: This involves measuring the dimensions of the vessel to ensure it meets the required specifications.
- Pressure testing: This involves testing the vessel's pressure capacity to ensure it is safe to use.

- Materials testing: This involves testing the materials used in the vessel to ensure they meet the required specifications.

What are the different types of pressure vessel fabrication?

Pressure vessel fabrication is the process of creating a pressure vessel using various materials and techniques. Some common types of pressure vessel fabrication include:

- Welding: This involves joining together different parts of the vessel using welding techniques.
- Bending: This involves shaping the vessel using bending techniques.
- Cutting: This involves cutting the vessel to the required size and shape using cutting techniques.
- Forming: This involves shaping the vessel using forming techniques, such as rolling or stretching.
- Machining: This involves machining the vessel to the required size and shape using machining techniques, such as turning or milling.

What are the different types of pressure vessel installation?

Pressure vessel installation is the process of installing the vessel in its intended location and preparing it for operation. Some common types of pressure vessel installation include:

- Setting: This involves placing the vessel in its intended location and securing it in place.
- Anchoring: This involves securing the vessel in place using anchors or other restraints.
- Piping: This involves connecting the vessel to other equipment or piping using connectors or fittings.
- Testing: This involves testing the vessel to ensure it is safe to use and meets the required specifications.
- Commissioning: This involves preparing the vessel for operation and performing any necessary calibrations or adjustments.

What are the different types of pressure vessel failure modes?

Pressure vessel failure can occur due to various factors, such as corrosion, fatigue, overloading, or improper design or fabrication. Some common types of pressure vessel failure modes include:

- Burst failure: This is a type of failure that occurs when the vessel ruptures or splits due to excessive internal pressure.
- Leakage failure: This is a type of failure that occurs when the vessel leaks due to a breach in its integrity.
- Collapse failure: This is a type of failure that occurs when the vessel collapses or buckles due to excessive internal pressure or external loads.
- Creep failure: This is a type of failure that occurs when the vessel deforms over time due to prolonged exposure to high temperatures or pressures.
- Fatigue failure: This is a type of failure that occurs when the vessel experiences repeated loading and unloading cycles, leading to cracks or fractures.

What are the different types of pressure vessel corrosion?

Pressure vessel corrosion is the process of deterioration of the vessel due to chemical reactions with the surrounding environment. Some common types of pressure vessel corrosion include:

- General corrosion: This is a type of corrosion that occurs uniformly over the surface of the vessel.
- Pitting corrosion: This is a type of corrosion that occurs in isolated areas of the vessel, forming small pits or holes.
- Galvanic corrosion: This is a type of corrosion that occurs when two different metals are in contact with each other in the presence of an electrolyte.
- Crevice corrosion: This is a type of corrosion that occurs in confined spaces or crevices of the vessel, such as under bolts or gaskets.

- Stress corrosion cracking: This is a type of corrosion that occurs when the vessel is subjected to both tensile stress and corrosive conditions.

What are the different types of pressure vessel materials of construction?

Pressure vessels can be made of various materials, depending on the intended use and the pressure and temperature conditions. Some common materials of construction for pressure vessels include:

- Carbon steel: This is a common material used for pressure vessels and is strong and durable.
- Stainless steel: This is a corrosion-resistant material that is commonly used for pressure vessels.
- Aluminum: This is a lightweight material that is commonly used for pressure vessels.
- Nickel alloys: These are corrosion-resistant materials that are commonly used for pressure vessels.
- Copper alloys: These are corrosion-resistant materials that are commonly used for pressure vessels.

What are the different types of pressure vessel sizes?

Pressure vessels can be made in various sizes, depending on the intended use and the capacity required. Some common sizes of pressure vessels include:

- Small pressure vessels: These are pressure vessels with a capacity of less than 50 liters.
- Medium pressure vessels: These are pressure vessels with a capacity of 50 to 1000 liters.
- Large pressure vessels: These are pressure vessels with a capacity of more than 1000 liters and more as per requirement and types of material used.

What are the different types of pressure vessel shapes?

Pressure vessels can be made in various shapes, depending on the intended use and the space constraints. Some common shapes of pressure vessels include:

- Cylindrical pressure vessels: These are pressure vessels with a cylindrical shape.
- Spherical pressure vessels: These are pressure vessels with a spherical shape.
- Conical pressure vessels: These are pressure vessels with a conical shape.
- Rectangular pressure vessels: These are pressure vessels with a rectangular shape.
- Irregular pressure vessels: These are pressure vessels with a custom or irregular shape.

What are the different types of pressure vessel regulations?

Pressure vessels are subject to various regulations and codes to ensure their safety and performance. Some common types of pressure vessel regulations include:

- ASME Boiler and Pressure Vessel Code: This is a widely used code that provides requirements for the design, fabrication, inspection, and testing of boilers and pressure vessels.
- ISO 4126: This is an international standard that provides guidelines for the design, fabrication, inspection, and testing of pressure vessels.
- PED (Pressure Equipment Directive): This is a European Union directive that provides requirements for the design, fabrication, inspection, and testing of pressure vessels.
- Australian Pressure Vessel Code: This is an Australian standard that provides requirements for the design, fabrication, inspection, and testing of pressure vessels.
- JIS B 8241: This is a Japanese standard that provides requirements for the design, fabrication, inspection, and testing

of pressure vessels.

What are the different types of pressure vessel design codes?

Pressure vessel design codes are used to provide guidelines for the design of pressure vessels to ensure their safety and performance. Some common types of pressure vessel design codes include:

- ASME Boiler and Pressure Vessel Code: This is a widely used code that provides requirements for the design of boilers and pressure vessels.
- ISO 4126: This is an international standard that provides guidelines for the design of pressure vessels.
- PED (Pressure Equipment Directive): This is a European Union directive that provides requirements for the design of pressure vessels.
- Australian Pressure Vessel Code: This is an Australian standard that provides requirements for the design of pressure vessels.
- JIS B 8241: This is a Japanese standard that provides requirements for the design of pressure vessels.

What are the different types of pressure vessel design calculations?

Pressure vessel design calculations are used to determine the size, shape, and thickness of the vessel based on the intended use and the pressure and temperature conditions. Some common types of pressure vessel design calculations include:

- Stress analysis: This involves calculating the stresses on the vessel due to internal pressure and external loads.
- Thickness calculation: This involves calculating the minimum required thickness of the vessel based on the stress analysis.
- Size calculation: This involves calculating the required size of the vessel based on the volume and pressure capacity needed.

- Shape calculation: This involves determining the optimal shape of the vessel based on the intended use and the space constraints.
- Material selection: This involves selecting the appropriate material for the vessel based on the pressure and temperature conditions and the corrosion resistance requirements.

What are the different types of pressure vessel design software?

Pressure vessel design software is used to assist in the design and analysis of pressure vessels. Some common types of pressure vessel design software include:

- PV Elite: This is a software tool that is widely used for the design and analysis of pressure vessels.
- CAESAR II: This is a software tool that is commonly used for the analysis of piping and pressure vessels.
- ASME BPVC Section VIII Division 2: This is a software tool that is used for the design of pressure vessels based on the ASME Boiler and Pressure Vessel Code.
- Compress: This is a software tool that is commonly used for the design and analysis of compressors and pressure vessels.
- ANSYS: This is a software tool that is commonly used for the analysis of structural and mechanical systems, including pressure vessels.

What are the different types of pressure vessel design factors?

Pressure vessel design involves considering various factors to ensure the safety and performance of the vessel. Some common design factors for pressure vessels include:

- Operating pressure: The maximum operating pressure of the vessel must be considered to ensure the vessel can withstand the internal pressure without failing.
- Operating temperature: The maximum operating temperature of the vessel must be considered to ensure the vessel can withstand

the temperature without failing or deforming.

- Material properties: The properties of the materials used in the vessel, such as strength, stiffness, and corrosion resistance, must be considered in the design.
- Loadings: The external loads that the vessel may be subjected to, such as wind or seismic loads, must be considered in the design.
- Geometry: The size and shape of the vessel must be considered based on the volume and pressure capacity needed and the space constraints.
- Corrosion: The corrosion resistance of the materials used in the vessel must be considered to ensure the vessel has a long service life.

What are the different types of pressure vessel design standards?

Pressure vessel design standards provide guidelines for the design of pressure vessels to ensure their safety and performance. Some common pressure vessel design standards include:

- ASME Boiler and Pressure Vessel Code: This is a widely used standard that provides requirements for the design of boilers and pressure vessels.
- ISO 4126: This is an international standard that provides guidelines for the design of pressure vessels.
- PED (Pressure Equipment Directive): This is a European Union directive that provides requirements for the design of pressure vessels.
- Australian Pressure Vessel Code: This is an Australian standard that provides requirements for the design of pressure vessels.
- JIS B 8241: This is a Japanese standard that provides requirements for the design of pressure vessels.

What are the different types of pressure vessel design considerations?

Pressure vessel design involves considering various factors to ensure the safety and performance of the vessel. Some common design considerations for pressure vessels include:

- Operating pressure: The maximum operating pressure of the vessel must be considered to ensure the vessel can withstand the internal pressure without failing.
- Operating temperature: The maximum operating temperature of the vessel must be considered to ensure the vessel can withstand the temperature without failing or deforming.
- Material selection: The appropriate material for the vessel must be selected based on the pressure and temperature conditions and the corrosion resistance requirements.
- Geometry: The size and shape of the vessel must be considered based on the volume and pressure capacity needed and the space constraints.
- Loadings: The external loads that the vessel may be subjected to, such as wind or seismic loads, must be considered in the design.
- Corrosion: The corrosion resistance of the materials used in the vessel must be considered to ensure the vessel has a long service life.
- Manufacturing processes: The manufacturing processes used to fabricate the vessel must be considered to ensure the vessel meets the required tolerances and specifications.

What are the different types of pressure vessel design calculations?

Pressure vessel design calculations are used to determine the size, shape, and thickness of the vessel based on the intended use and the pressure and temperature conditions. Some common types of pressure vessel design calculations include:

Stress analysis: This involves calculating the stresses on the vessel due to internal pressure and external loads.

Thickness calculation: This involves calculating the minimum required thickness of the vessel based on the stress analysis.

Size calculation: This involves calculating the required size of the vessel based on the volume and pressure capacity needed.

Shape calculation: This involves determining the optimal shape of the vessel based on the intended use and the space constraints.

Material selection: This involves selecting the appropriate material for the vessel based on the pressure and temperature conditions and the corrosion resistance requirements.

What are the different types of pressure vessel fabrication processes?

Pressure vessel fabrication involves the process of constructing the vessel from raw materials according to the design specifications. Some common types of pressure vessel fabrication processes include:

- Cutting: This involves cutting the materials to the required shape and size using methods such as sawing, shearing, or laser cutting.
- Forming: This involves shaping the materials using methods such as rolling, bending, or forging.
- Welding: This involves joining the materials using welding techniques, such as shielded metal arc welding (SMAW), gas metal arc welding (GMAW), or tungsten inert gas welding (TIG).
- Heat treatment: This involves subjecting the materials to heat and cooling cycles to improve their properties, such as strength or ductility.
- Machining: This involves removing material from the surface of the materials using machining techniques, such as turning or milling.

What are the different types of pressure vessel fabrication methods?

Pressure vessel fabrication involves the process of constructing the vessel from raw materials according to the design specifications. Some common types of pressure vessel fabrication methods include:

- Welded fabrication: This involves constructing the vessel from welded components.
- Bolted fabrication: This involves constructing the vessel from bolted components.
- Riveted fabrication: This involves constructing the vessel from riveted components.
- Cladded fabrication: This involves constructing the vessel from cladded materials, where a layer of one material is bonded to the surface of another material.
- Forged fabrication: This involves constructing the vessel from forged components, which are formed by shaping the material using high temperatures and pressures.

What are the different types of pressure vessel inspection methods?

Pressure vessel inspections are used to evaluate the safety and performance of the vessel and ensure that it meets the required specifications. Some common types of pressure vessel inspection methods include:

- Visual inspection: This involves visually inspecting the vessel for any signs of damage or wear.
- Non-destructive testing: This involves using testing methods, such as radiographic testing or ultrasonic testing, to inspect the vessel for defects or imperfections.
- Dimensional inspection: This involves measuring the dimensions of the vessel to ensure they meet the required tolerances and specifications.
- Hydrostatic testing: This involves pressurizing the vessel with water to test its strength and integrity.
- Pneumatic testing: This involves pressurizing the vessel with air to test its strength and integrity.

CHAPTER THREE

Questions and answers related to codes and standards for pressure vessels

Here are some questions and answers related to codes and standards for pressure vessels:

What codes and standards apply to the design, fabrication, and inspection of pressure vessels?

Some commonly used codes and standards for pressure vessels include:

- ASME Boiler and Pressure Vessel Code (BPVC).
- API 510: Pressure Vessel Inspection Code: In-Service Inspection, Rating, Repair, and Alteration.
- API 579-1/ASME FFS-1: Fitness-For-Service.
- EN 13445: Unfired Pressure Vessels.
- ISO 4706: Unfired pressure vessels - Acceptance inspection.

What is the ASME Boiler and Pressure Vessel Code (BPVC)?

The ASME Boiler and Pressure Vessel Code (BPVC) is a set of rules for the design, construction, and inspection of boilers and pressure vessels. It is published by the American Society of

Mechanical Engineers (ASME).

What is the API 510: Pressure Vessel Inspection Code?

API 510: Pressure Vessel Inspection Code is a standard published by the American Petroleum Institute (API) that provides guidelines for the in-service inspection, rating, repair, and alteration of pressure vessels.

What is the API 579-1/ASME FFS-1: Fitness-For-Service standard?

API 579-1/ASME FFS-1: Fitness-For-Service is a standard that provides guidance on the assessment and repair of existing pressure vessels and piping systems. It is intended to help determine the appropriate repair or modification for equipment that is damaged or degraded, but still in service.

What is the EN 13445: Unfired Pressure Vessels standard?

EN 13445: Unfired Pressure Vessels is a European standard for the design and construction of unfired pressure vessels. It applies to pressure vessels made of metallic materials, and covers vessels that are intended to be used at ambient temperature or at a temperature that does not exceed 450°C.

What is the ISO 4706: Unfired pressure vessels - Acceptance inspection standard?

ISO 4706: Unfired pressure vessels - Acceptance inspection is an international standard that provides guidelines for the acceptance inspection of unfired pressure vessels. It covers the inspection of materials, fabrication, and testing of pressure vessels.

What is the scope of the ASME BPVC?

The ASME BPVC covers the design, construction, and inspection of boilers and pressure vessels. It includes rules for materials, design, fabrication, examination, inspection, testing, and certification of boilers and pressure vessels.

What is the difference between the ASME BPVC and the API 510: Pressure Vessel Inspection Code?

The ASME BPVC is a code for the design, construction, and inspection of boilers and pressure vessels, while the API 510: Pressure Vessel Inspection Code is a standard for the in-service inspection, rating, repair, and alteration of pressure vessels. The ASME BPVC covers the initial design and construction of pressure vessels, while the API 510 covers the ongoing maintenance and repair of in-service pressure vessels.

How often should pressure vessels be inspected according to the API 510: Pressure Vessel Inspection Code?

The frequency of inspections for pressure vessels is based on the vessel's service, size, and operating conditions. The API 510: Pressure Vessel Inspection Code provides guidelines for determining the appropriate inspection interval for a given vessel. It is important to follow these guidelines to ensure the safe and reliable operation of pressure vessels.

Can pressure vessels be repaired and modified according to the ASME BPVC?

Yes, the ASME BPVC includes rules for the repair and modification of pressure vessels. These rules cover the evaluation and repair of damaged or degraded vessels, as well as the modification of vessels to change their design or operating conditions.

Can pressure vessels be used at temperatures above 450°C according to the EN 13445: Unfired Pressure Vessels standard?

No, the EN 13445: Unfired Pressure Vessels standard only applies to pressure vessels that are intended to be used at ambient temperature or at a temperature that does not exceed 450°C. For pressure vessels that will be used at higher temperatures, a different standard may be applicable.

Is the ISO 4706: Unfired pressure vessels - Acceptance inspection standard only applicable to metallic pressure vessels?

Yes, the ISO 4706: Unfired pressure vessels - Acceptance inspection standard only covers the inspection of unfired pressure vessels made of metallic materials. Non-metallic pressure vessels, such as those made of plastic or composite materials, may be covered by different standards.

Are the rules in the ASME BPVC mandatory?

The rules in the ASME BPVC are voluntary, but they are widely used and recognized as the industry standard for the design and construction of boilers and pressure vessels. Many local, state, and federal agencies have adopted the ASME BPVC as a mandatory standard, and compliance with the code is often required to obtain permits and approvals for the construction and operation of boilers and pressure vessels.

Can the ASME BPVC be used for pressure vessels used in international projects?

Yes, the ASME BPVC is a widely recognized and respected standard that is used for the design and construction of boilers and pressure vessels around the world. It is commonly used for international projects, and is recognized by many countries as an acceptable standard for the design and construction of pressure vessels.

Is the API 510: Pressure Vessel Inspection Code only applicable to pressure vessels in the oil and gas industry?

No, the API 510: Pressure Vessel Inspection Code is not limited to the oil and gas industry. It is a widely used standard that applies to pressure vessels in a variety of industries, including chemical processing, power generation, and manufacturing.

Is the API 579-1/ASME FFS-1: Fitness-For-Service standard only applicable to pressure vessels made of metallic materials?

No, the API 579-1/ASME FFS-1: Fitness-For-Service standard can be used to assess the fitness-for-service of pressure vessels made of metallic and non-metallic materials. It provides guidelines for evaluating the remaining service life of damaged or degraded pressure vessels and piping systems, and can be used to determine the appropriate repair or modification for a given vessel.

Are the rules in the EN 13445: Unfired Pressure Vessels standard mandatory in all European countries?

No, the EN 13445: Unfired Pressure Vessels standard is a voluntary European standard that is not legally binding in all European countries. However, it is widely used and recognized as a good practice standard for the design and construction of unfired pressure vessels in Europe.

Is the ISO 4706: Unfired pressure vessels - Acceptance inspection standard only applicable to new pressure vessels?

No, the ISO 4706: Unfired pressure vessels - Acceptance inspection standard can be used for both new and existing pressure vessels. It provides guidelines for the inspection of materials, fabrication, and testing of pressure vessels, and can be used to ensure that new vessels meet the required specifications and that existing vessels are in good condition.

What is the role of an authorized inspector in the design and construction of pressure vessels according to the ASME BPVC?

An authorized inspector is an individual who is qualified to witness and certify the construction of pressure vessels according to the ASME BPVC. The authorized inspector is responsible for reviewing the design and construction of the vessel to ensure that it meets the requirements of the code, and for issuing a certificate of compliance upon completion.

What is the role of an API 510: Pressure Vessel Inspection Code authorized inspector?

An API 510: Pressure Vessel Inspection Code authorized inspector is a person who is qualified to conduct inspections of in-service pressure vessels according to the API 510 standard. The authorized inspector is responsible for evaluating the condition of the vessel, determining the appropriate inspection interval, and verifying that the vessel meets the requirements of the code.

Can the API 579-1/ASME FFS-1: Fitness-For-Service standard be used to assess the fitness-for-service of pressure vessels made of non-metallic materials?

Yes, the API 579-1/ASME FFS-1: Fitness-For-Service standard can be used to assess the fitness-for-service of pressure vessels made of both metallic and non-metallic materials. It provides guidelines for evaluating the remaining service life of damaged or degraded pressure vessels and piping systems, and can be used to determine the appropriate repair or modification for a given vessel, regardless of its material.

Can the EN 13445: Unfired Pressure Vessels standard be used for the design and construction of fired pressure vessels?

No, the EN 13445: Unfired Pressure Vessels standard only applies to unfired pressure vessels. For fired pressure vessels, such as boilers, a different standard may be applicable.

Is the ISO 4706: Unfired pressure vessels - Acceptance inspection standard only applicable to pressure vessels used in the process industry?

No, the ISO 4706: Unfired pressure vessels - Acceptance inspection standard can be used for pressure vessels in a variety of industries, including the process industry, oil and gas, and power generation. It provides guidelines for the inspection of materials, fabrication, and testing of pressure vessels, and can be used to ensure that the vessels meet the required specifications and are in good condition.

Can the ASME BPVC be used for the design and construction of pressure vessels made of non-metallic materials?

Yes, the ASME BPVC includes rules for the design and construction of pressure vessels made of non-metallic materials, such as plastic and composite materials. These rules are included in the ASME BPVC Division 3: Construction of High Pressure Vessels.

Can the API 510: Pressure Vessel Inspection Code be used to inspect pressure vessels made of non-metallic materials?

Yes, the API 510: Pressure Vessel Inspection Code can be used to inspect pressure vessels made of both metallic and non-metallic materials. It provides guidelines for the in-service inspection of pressure vessels, and can be used to evaluate the condition of the vessel and determine the appropriate inspection interval.

Can the API 579-1/ASME FFS-1: Fitness-For-Service standard be used to determine the remaining service life of pressure vessels made of non-metallic materials?

Yes, the API 579-1/ASME FFS-1: Fitness-For-Service standard can be used to determine the remaining service life of pressure vessels made of both metallic and non-metallic materials. It provides guidelines for evaluating the condition of damaged or degraded pressure vessels and piping systems, and can be used to determine the appropriate repair or modification for a given vessel, regardless of its material.

Does the EN 13445: Unfired Pressure Vessels standard cover the inspection of pressure vessels?

No, the EN 13445: Unfired Pressure Vessels standard is primarily a design and construction standard, and it does not cover the inspection of pressure vessels. For inspection guidelines, a different standard, such as the ISO 4706: Unfired pressure vessels - Acceptance inspection standard, may be applicable.

Can the ISO 4706: Unfired pressure vessels - Acceptance inspection standard be used to inspect pressure vessels made of non-metallic materials?

Yes, the ISO 4706: Unfired pressure vessels - Acceptance inspection standard can be used to inspect pressure vessels made of both metallic and non-metallic materials. It provides guidelines for the inspection of materials, fabrication, and testing of pressure vessels, and can be used to ensure that the vessels meet the required specifications and are in good condition, regardless of their material.

What is the ASME "U" stamp?

The ASME "U" stamp is a symbol of certification indicating that a pressure vessel has been designed and constructed in accordance with the requirements of the ASME BPVC. The "U" stamp is issued by the ASME and is only granted to pressure vessels that have been fabricated by an ASME-certified manufacturer and inspected by an authorized inspector.

What is the ASME "R" stamp?

The ASME "R" stamp is a symbol of certification indicating that a pressure vessel has been repaired or altered in accordance with the requirements of the ASME BPVC. The "R" stamp is issued by the ASME and is only granted to pressure vessels that have been repaired or altered by an ASME-certified manufacturer and inspected by an authorized inspector.

Can the ASME "U" stamp be used on pressure vessels made of non-metallic materials?

Yes, the ASME "U" stamp can be used on pressure vessels made of both metallic and non-metallic materials. The ASME BPVC includes rules for the design and construction of pressure vessels made of non-metallic materials, and these vessels are eligible for the "U" stamp if they meet the requirements of the code.

Can the ASME "R" stamp be used on pressure vessels made of non-metallic materials?

Yes, the ASME "R" stamp can be used on pressure vessels made of both metallic and non-metallic materials. The ASME BPVC includes rules for the repair and alteration of pressure vessels made of non-metallic materials, and these vessels are eligible for the "R" stamp if they meet the requirements of the code.

Is the ASME "U" stamp required for all pressure vessels in the United States?

No, the ASME "U" stamp is not required for all pressure vessels in the United States. It is a voluntary certification that is granted to pressure vessels that meet the requirements of the ASME BPVC. However, many local, state, and federal agencies have adopted the ASME BPVC as a mandatory standard, and compliance with the code is often required to obtain permits and approvals for the construction and operation of boilers and pressure vessels.

Is the ASME "R" stamp required for all pressure vessels in the United States?

No, the ASME "R" stamp is not required for all pressure vessels in the United States. It is a voluntary certification that is granted to pressure vessels that have been repaired or altered in accordance with the requirements of the ASME BPVC. However, many local, state, and federal agencies have adopted the ASME BPVC as a mandatory standard, and compliance with the code is often required to obtain permits and approvals for the repair and alteration of boilers and pressure vessels.

What is the ASME "PP" stamp?

The ASME "PP" stamp is a symbol of certification indicating that a pressure vessel has been designed and constructed in accordance with the requirements of the ASME BPVC for use in the petrochemical and petroleum industries. The "PP" stamp is issued by the ASME and is only granted to pressure vessels that have been fabricated by an ASME-certified manufacturer and inspected by an authorized inspector.

What is the ASME "H" stamp?

The ASME "H" stamp is a symbol of certification indicating that a pressure vessel has been designed and constructed in accordance with the requirements of the ASME BPVC for use in the hydrogen service. The "H" stamp is issued by the ASME and is only granted to pressure vessels that have been fabricated by an ASME-certified manufacturer and inspected by an authorized inspector.

What is the ASME "NB" stamp?

The ASME "NB" stamp is a symbol of certification indicating that a pressure vessel has been designed and constructed in accordance with the requirements of the ASME BPVC for use in the nuclear power industry. The "NB" stamp is issued by the ASME and is only granted to pressure vessels that have been fabricated by an ASME-certified manufacturer and inspected by an authorized inspector.

Are the ASME "PP", "H", and "NB" stamps required for all pressure vessels in the petrochemical, hydrogen, and nuclear power industries?

No, the ASME "PP", "H", and "NB" stamps are not required for all pressure vessels in these industries. They are voluntary certifications that are granted to pressure vessels that meet the requirements of the ASME BPVC for specific service conditions. However, many companies in these industries choose to use pressure vessels with these stamps as a way to demonstrate the quality and safety of their equipment.

Can the ASME "PP", "H", and "NB" stamps be used on pressure vessels made of non-metallic materials?

Yes, the ASME "PP", "H", and "NB" stamps can be used on pressure vessels made of both metallic and non-metallic materials.

The ASME BPVC includes rules for the design and construction of pressure vessels made of non-metallic materials, and these vessels are eligible for these stamps if they meet the requirements of the code for the specific service conditions.

What is the API "V" stamp?

The API "V" stamp is a symbol of certification indicating that a pressure vessel has been designed and constructed in accordance with the requirements of the API 510: Pressure Vessel Inspection Code. The "V" stamp is issued by the American Petroleum Institute (API) and is only granted to pressure vessels that have been fabricated by an API-certified manufacturer and inspected by an authorized inspector.

What is the API "R" stamp?

The API "R" stamp is a symbol of certification indicating that a pressure vessel has been repaired or altered in accordance with the requirements of the API 510: Pressure Vessel Inspection Code. The "R" stamp is issued by the American Petroleum Institute (API) and is only granted to pressure vessels that have been repaired or altered by an API-certified manufacturer and inspected by an authorized inspector.

Can the API "V" stamp be used on pressure vessels made of non-metallic materials?

Yes, the API "V" stamp can be used on pressure vessels made of both metallic and non-metallic materials. The API 510: Pressure Vessel Inspection Code includes rules for the design and construction of pressure vessels made of non-metallic materials, and these vessels are eligible for the "V" stamp if they meet the requirements of the code.

Can the API "R" stamp be used on pressure vessels made of non-metallic materials?

Yes, the API "R" stamp can be used on pressure vessels made of both metallic and non-metallic materials. The API 510: Pressure Vessel Inspection Code includes rules for the repair and alteration of pressure vessels made of non-metallic materials, and these vessels are eligible for the "R" stamp if they meet the requirements of the code.

Is the API "V" stamp required for all pressure vessels in the United States?

No, the API "V" stamp is not required for all pressure vessels in the United States. It is a voluntary certification that is granted to pressure vessels that meet the requirements of the API 510: Pressure Vessel Inspection Code. However, many companies in the oil and gas industry choose to use pressure vessels with the "V" stamp as a way to demonstrate the quality and safety of their equipment.

Is the API "R" stamp required for all pressure vessels in the United States?

No, the API "R" stamp is not required for all pressure vessels in the United States. It is a voluntary certification that is granted to pressure vessels that have been repaired or altered in accordance with the requirements of the API 510: Pressure Vessel Inspection Code. However, many companies in the oil and gas industry choose to use pressure vessels with the "R" stamp as a way to demonstrate the quality and safety of their equipment.

What is the PED?

The PED (Pressure Equipment Directive) is a European Union directive that sets out the safety requirements for the design, manufacture, and testing of pressure vessels and other pressure equipment. The PED applies to all pressure equipment with a maximum allowable pressure greater than 0.5 bar, and is intended to ensure that pressure equipment is safe for use within the EU.

Is the PED mandatory in all European countries?

Yes, the PED is mandatory in all EU member states, as well as in certain other European countries that have adopted the directive. The PED sets out the minimum safety requirements for pressure equipment within the EU, and compliance with the directive is required in order to place pressure equipment on the market or put it into service within the EU.

Can the PED be used for the design and construction of pressure vessels made of non-metallic materials?

Yes, the PED can be used for the design and construction of pressure vessels made of both metallic and non-metallic materials. The directive includes specific requirements for the design and testing of pressure vessels made of non-metallic materials, and these vessels must meet these requirements in order to be compliant with the PED.

Can the PED be used for the inspection of pressure vessels?

Yes, the PED includes provisions for the periodic inspection of pressure vessels and other pressure equipment. The directive requires that pressure equipment be inspected at regular intervals in order to ensure that it remains safe for use, and it sets out the minimum requirements for the inspection and testing of pressure equipment.

Is the PED equivalent to the ASME BPVC?

No, the PED and the ASME BPVC are two separate standards with different requirements and scope. While both standards address the design and construction of pressure vessels, they are not directly equivalent and compliance with one standard does not necessarily guarantee compliance with the other.

Can the PED and the ASME BPVC be used together for the design and construction of pressure vessels?

Yes, it is possible to use both the PED and the ASME BPVC together for the design and construction of pressure vessels. However, it is important to carefully review the requirements of both standards and ensure that the vessel meets the applicable requirements of each.

What is the CE mark?

The CE mark is a symbol that indicates that a product meets the safety, health, and environmental protection requirements of the European Union. The CE mark is required for a wide range of products, including pressure vessels, and is intended to ensure that these products are safe for use within the EU.

Is the CE mark required for all pressure vessels in the European Union?

Yes, the CE mark is required for all pressure vessels placed on the market or put into service within the EU. The CE mark indicates that the pressure vessel has been designed and manufactured in accordance with the applicable EU directives, including the PED, and that it meets the minimum safety requirements for use within the EU.

Can the CE mark be used for pressure vessels made of non-metallic materials?

Yes, the CE mark can be used for pressure vessels made of both metallic and non-metallic materials. The CE mark applies to all pressure vessels, regardless of their material, and is intended to ensure that these vessels are safe for use within the EU.

What is the CRN?

The CRN (Canadian Registration Number) is a symbol of certification indicating that a pressure vessel has been designed and constructed in accordance with the requirements of the Canadian Boiler, Pressure Vessel, and Pressure Piping Code. The CRN is issued by the Canadian Registration Authorities (CRAs) and is only granted to pressure vessels that have been fabricated by a CRA-certified manufacturer and inspected by an authorized inspector.

Is the CRN required for all pressure vessels in Canada?

No, the CRN is not required for all pressure vessels in Canada. It is a voluntary certification that is granted to pressure vessels that meet the requirements of the Canadian Boiler, Pressure Vessel, and Pressure Piping Code. However, many local, provincial, and federal agencies have adopted the code as a mandatory standard, and compliance with the code is often required to obtain permits and approvals for the construction and operation of boilers and pressure vessels.

Can the CRN be used for pressure vessels made of non-metallic materials?

Yes, the CRN can be used for pressure vessels made of both metallic and non-metallic materials. The Canadian Boiler, Pressure Vessel, and Pressure Piping Code includes rules for the design and construction of pressure vessels made of non-metallic materials, and these vessels are eligible for the CRN if they meet the requirements of the code.

What is the R stamp?

The R stamp is a symbol of certification indicating that a pressure vessel has been repaired or altered in accordance with the requirements of the ASME BPVC. The R stamp is issued by the ASME and is only granted to pressure vessels that have been repaired or altered by an ASME-certified manufacturer and inspected by an authorized inspector. The R stamp is used in conjunction with the U stamp, which indicates that the vessel has been designed and constructed in accordance with the ASME BPVC.

What is the EN 13445: Unfired Pressure Vessels standard?

The EN 13445: Unfired Pressure Vessels standard is a European standard that specifies the requirements for the design, manufacture, inspection, and testing of unfired pressure vessels made of metallic materials. The standard covers vessels used in a variety of industries, including chemical, petrochemical, pharmaceutical, and food processing.

Is the EN 13445: Unfired Pressure Vessels standard mandatory in all European countries?

No, the EN 13445: Unfired Pressure Vessels standard is not mandatory in all European countries. It is a voluntary standard that provides guidelines for the design, manufacture, inspection, and testing of unfired pressure vessels. However, many countries in Europe have adopted the standard as a national standard, and compliance with the EN 13445 is often required to obtain permits and approvals for the construction and operation of pressure vessels.

Can the EN 13445: Unfired Pressure Vessels standard be used for pressure vessels made of non-metallic materials?

No, the EN 13445: Unfired Pressure Vessels standard is specifically for pressure vessels made of metallic materials. For pressure vessels made of non-metallic materials, a different standard, such as the ISO 14122: Safety of machinery - Permanent means of access to machinery may be applicable.

What is the ISO 4706: Unfired pressure vessels - Acceptance inspection standard?

The ISO 4706: Unfired pressure vessels - Acceptance inspection standard is an international standard that provides guidelines for the inspection of materials, fabrication, and testing of unfired pressure vessels. The standard covers vessels made of both metallic and non-metallic materials and is intended to ensure that these vessels meet the required specifications and are in good condition.

Is the ISO 4706: Unfired pressure vessels - Acceptance inspection standard mandatory?

No, the ISO 4706: Unfired pressure vessels - Acceptance inspection standard is not a mandatory standard. It is a voluntary standard that provides guidelines for the inspection of unfired pressure vessels. However, many companies choose to use the standard as a way to ensure the quality and safety of their pressure vessels.

What is the API RP 579: Fitness-For-Service standard?

The API RP 579: Fitness-For-Service standard is a recommended practice developed by the American Petroleum Institute (API) that provides guidelines for the evaluation of pressure vessels and other equipment for continued service. The standard covers a variety of equipment, including pressure vessels, piping, tanks, and heat exchangers, and is intended to help operators determine the fitness-for-service of their equipment.

Is the API RP 579: Fitness-For-Service standard mandatory?

No, the API RP 579: Fitness-For-Service standard is not a mandatory standard. It is a recommended practice that provides guidelines for the evaluation of equipment for continued service. However, many companies in the oil and gas industry choose to use the standard as a way to ensure the safety and reliability of their equipment.

What is the API RP 581: Risk-Based Inspection standard?

The API RP 581: Risk-Based Inspection standard is a recommended practice developed by the American Petroleum Institute (API) that provides guidelines for the implementation of a risk-based inspection (RBI) program for pressure vessels and other equipment. The standard covers a variety of equipment, including pressure vessels, piping, tanks, and heat exchangers, and is intended to help operators prioritize their inspection efforts based on the risk of failure.

Is the API RP 581: Risk-Based Inspection standard mandatory?

No, the API RP 581: Risk-Based Inspection standard is not a mandatory standard. It is a recommended practice that provides guidelines for the implementation of a risk-based inspection program. However, many companies in the oil and gas industry choose to use the standard as a way to ensure the safety and reliability of their equipment.

What is the ISO 10418: Petroleum and natural gas industries - Design and operation of subsea production systems standard?

The ISO 10418: Petroleum and natural gas industries - Design and operation of subsea production systems standard is an international standard that provides guidelines for the design, manufacture, installation, and operation of subsea production systems. The standard covers a variety of equipment, including pressure vessels, piping, and control systems, and is intended to ensure the safety and reliability of subsea production systems.

What is the ISO 10419: Petroleum and natural gas industries - Subsea production control systems standard?

The ISO 10419: Petroleum and natural gas industries - Subsea production control systems standard is an international standard that provides guidelines for the design, manufacture, installation, and operation of subsea production control systems. The standard covers a variety of equipment, including pressure vessels, piping, and control systems, and is intended to ensure the safety and reliability of subsea production control systems.

What is the ISO 10423: Petroleum and natural gas industries - Drilling and production equipment - Subsea wellhead and Christmas tree equipment standard?

The ISO 10423: Petroleum and natural gas industries - Drilling and production equipment - Subsea wellhead and Christmas tree equipment standard is an international standard that provides guidelines for the design, manufacture, installation, and operation of subsea wellhead and Christmas tree equipment. The standard covers a variety of equipment, including pressure vessels, piping, and control systems, and is intended to ensure the safety and reliability of subsea wellhead and Christmas tree equipment.

What is the ISO 10424: Petroleum and natural gas industries - Drilling and production equipment - Subsea wellhead and Christmas tree equipment - Part 2: Testing standard?

The ISO 10424: Petroleum and natural gas industries - Drilling and production equipment - Subsea wellhead and Christmas tree equipment - Part 2: Testing standard is an international standard that provides guidelines for the testing of subsea wellhead and Christmas tree equipment. The standard covers a variety of equipment, including pressure vessels, piping, and control systems, and is intended to ensure that these systems are fit for service.

What is the ISO 10425: Petroleum and natural gas industries - Drilling and production equipment - Subsea wellhead and Christmas tree equipment - Part 3: Inspection and maintenance standard?

The ISO 10425: Petroleum and natural gas industries - Drilling and production equipment - Subsea wellhead and Christmas tree equipment - Part 3: Inspection and maintenance standard is an international standard that provides guidelines for the inspection and maintenance of subsea wellhead and Christmas tree equipment.

The standard covers a variety of equipment, including pressure vessels, piping, and control systems, and is intended to ensure the safety and reliability of these systems over their lifespan.

CHAPTER FOUR

Question and answer related to pressure vessel Design

Few more questions and answers related to pressure vessel design:

Question: How is the thickness of a pressure vessel determined?

Answer: The thickness of a pressure vessel is determined through the use of design pressure, design temperature, and materials of construction. The design pressure is the maximum pressure that the vessel is expected to experience during normal operation, and the design temperature is the maximum temperature that the vessel is expected to experience during normal operation. The materials of construction are used to determine the minimum required thickness of the vessel based on the design pressure and temperature using a pressure vessel design code, such as the ASME Boiler and Pressure Vessel Code. The thickness of the vessel may also be increased to account for corrosion, erosion, or other factors that may weaken the vessel over time.

Question: What is the purpose of a pressure vessel?

Answer: A pressure vessel is a container designed to hold gases or liquids at a pressure different from the ambient pressure. They are used in a variety of industries, such as chemical processing, oil and gas production, and power generation, for storing and transporting fluids under high pressure.

Question: What are some important considerations in the design of a pressure vessel?

Answer: Some important considerations in the design of a pressure vessel include the materials of construction, pressure and temperature ratings, structural integrity, corrosion resistance, and ease of maintenance. The materials of construction must be able to withstand the pressure and temperature of the contents, and also be compatible with the fluid being stored. The vessel must be structurally sound to avoid failure under operating conditions. It must also be resistant to corrosion to ensure it can operate safely for its intended lifespan. Ease of maintenance is also important, as it can help reduce downtime and prolong the vessel's operational life.

Question: What are some common codes and standards that apply to the design and fabrication of pressure vessels?

Answer: Some common codes and standards that apply to the design and fabrication of pressure vessels include the ASME Boiler and Pressure Vessel Code (ASME BPVC), the American Petroleum Institute (API) standards, and the European Pressure Equipment Directive (PED). These codes and standards provide requirements for the design, fabrication, inspection, and testing of pressure vessels to ensure they are safe and fit for their intended purpose.

Question: What is a pressure vessel head?

Answer: A pressure vessel head is a part of a pressure vessel that forms one end of the vessel and is designed to close off and seal the vessel. The head is typically located at one end of the vessel and is attached to the vessel body using flanges, bolts, or welds. There are several types of pressure vessel heads, such as elliptical heads, torispherical heads, and hemispherical heads, which have different shapes and are used for different applications. The head must be able to withstand the internal pressure of the vessel and also be compatible with the materials of construction.

Question: What is a pressure vessel nozzle?

Answer: A pressure vessel nozzle is a pipe connection that allows fluids to flow into or out of a pressure vessel. The nozzle is typically located on the vessel body and is used to attach piping or other equipment to the vessel. The size and orientation of the nozzle must be designed to accommodate the flow of fluid and the pressure and temperature ratings of the vessel. The nozzle must also be compatible with the materials of construction of the vessel and the piping or equipment being attached.

Question: What is a pressure vessel support system?

Answer: A pressure vessel support system is a structure that supports the weight of the pressure vessel and any additional loads, such as wind or seismic loads. The support system must be able to withstand these loads and maintain the stability of the vessel. The design of the support system depends on the size, shape, and weight of the vessel, as well as the location and type of loads that it will be subjected to. The support system may include foundations, pedestals, cradles, or other structural elements to provide the necessary support.

Question: What is a pressure vessel shell?

Answer: The pressure vessel shell is the main structural component of the vessel that is designed to withstand the internal pressure of the vessel. The shell is typically cylindrical in shape and is made of a strong, durable material, such as steel or aluminum. The thickness of the shell is determined by the design pressure and temperature of the vessel and the materials of construction. The shell is usually welded or bolted to the vessel heads and may also be reinforced with additional structural elements, such as rings or stiffeners.

Question: What is a pressure vessel joint?

Answer: A pressure vessel joint is a connection between two parts of the vessel that allows them to be securely fastened together. The joint must be able to withstand the internal pressure of the vessel and also be compatible with the materials of construction. There are several types of pressure vessel joints, such as flanged joints, butt welded joints, and socket welded joints, which are used for different applications. The joint must also be designed to allow for thermal expansion and contraction of the vessel.

Question: What is a pressure vessel inspection?

Answer: A pressure vessel inspection is a process of evaluating the condition and safety of a pressure vessel to ensure it is fit for service. The inspection may include visual inspection, non-destructive testing, and destructive testing, depending on the age and condition of the vessel. The inspection is typically performed by a qualified inspector who follows a set of established guidelines, such as the ASME Boiler and Pressure Vessel Code or other industry standards. The purpose of the inspection is to identify any deficiencies or defects that may affect the safety or performance of the vessel and recommend repairs or replacements as needed.

Question: What are some common types of pressure vessel failure?

Answer: Some common types of pressure vessel failure include corrosion, fatigue, and overpressure. Corrosion is the gradual deterioration of the vessel materials due to the action of the contained fluid or the environment. Fatigue is the failure of the vessel due to the repeated application of stress, such as from pressure fluctuations or thermal cycling. Overpressure is the failure of the vessel due to the internal pressure exceeding the design pressure. Other types of failure may include fractures, cracks, leaks, and deformations.

Question: How is the design pressure of a pressure vessel determined?

Answer: The design pressure of a pressure vessel is the maximum pressure that the vessel is expected to experience during normal operation. The design pressure is typically based on the maximum operating pressure, the maximum intended pressure, and any additional factors that may increase the pressure, such as the effects of temperature or pressure spikes. The design pressure must be sufficient to ensure the safety of the vessel and its contents, but it should not be unnecessarily high, as this can increase the size and cost of the vessel.

Question: What are some common materials used for pressure vessels?

Answer: Some common materials used for pressure vessels include carbon steel, stainless steel, aluminum, and fiberglass reinforced plastic (FRP). The selection of the material depends on the design pressure and temperature, the compatibility with the contained fluid, and the desired corrosion resistance. Carbon steel is a strong, durable material that is widely used for pressure vessels, but it may corrode in certain environments. Stainless steel is more resistant to corrosion, but it is more expensive and may not be suitable for high temperature applications. Aluminum is lightweight and has good corrosion resistance, but it may not be suitable for high pressure applications. FRP is a composite material that is lightweight, corrosion resistant, and has good structural integrity, but it may not be suitable for high temperature applications.

Question: How is the design temperature of a pressure vessel determined?

Answer: The design temperature of a pressure vessel is the maximum temperature that the vessel is expected to experience during normal operation. The design temperature is typically based on the maximum operating temperature, the maximum intended temperature, and any additional factors that may increase the temperature, such as the effects of heat transfer or chemical reactions. The design temperature must be sufficient to ensure the safety of the vessel and its contents, but it should not be unnecessarily high, as this can increase the size and cost of the vessel.

Question: What is a pressure vessel pressure relief valve?

Answer: A pressure vessel pressure relief valve is a safety device that is installed on a pressure vessel to protect it from overpressure. The valve is designed to open at a predetermined pressure and release excess pressure to prevent the vessel from exceeding its design pressure. The pressure relief valve must be carefully selected and sized to ensure it can handle the maximum expected flow rate and relieving capacity. The valve must also be properly installed and maintained to ensure it functions properly in the event of an overpressure situation.

Question: How is the design life of a pressure vessel determined?

Answer: The design life of a pressure vessel is the expected length of time that the vessel is expected to operate safely and reliably. The design life is typically based on the materials of construction, the design pressure and temperature, the corrosion resistance, and the maintenance schedule. The design life must be sufficient to meet the operational needs of the vessel, but it should not be unnecessarily long, as this can increase the size and cost of the vessel. The design life may be expressed in terms of the number of operating hours, the number of operating cycles, or the number of years of service.

Question: What is a pressure vessel flange?

Answer: A pressure vessel flange is a mechanical connector that is used to join two parts of a pressure vessel, such as a head and a shell, or a nozzle and a pipe. The flange consists of a flat, circular plate with bolt holes and is used to connect the parts using bolts or studs. There are several types of pressure vessel flanges, such as slip-on flanges, weld neck flanges, and threaded flanges, which are used for different applications. The flange must be compatible with the materials of construction of the vessel and the pressure and temperature ratings.

Question: What is a pressure vessel design code?

Answer: A pressure vessel design code is a set of guidelines and requirements that are used to design and construct safe and reliable pressure vessels. The design code provides rules for the materials of construction, the structural integrity, the fabrication and testing, and the inspection and certification of pressure vessels. Some common pressure vessel design codes include the ASME Boiler and Pressure Vessel Code (ASME BPVC), the American Petroleum Institute (API) standards, and the European Pressure Equipment Directive (PED). The design code may also specify the minimum required thickness of the vessel based on the design pressure and temperature.

Question: What is a pressure vessel stamp?

Answer: A pressure vessel stamp is a marking or symbol that is applied to a pressure vessel to indicate that it has been designed, fabricated, and tested in accordance with a specific design code or standard. The stamp typically includes the name of the design code, the symbol of the code, the vessel's serial number, and the name and location of the manufacturer. The stamp is usually applied to the vessel using a stamping die or an embossing tool and is located in a visible location on the vessel. The stamp is used to identify the vessel and to verify that it has been properly constructed and tested.

Question: What is a pressure vessel hydrostatic test?

Answer: A pressure vessel hydrostatic test is a procedure that is used to test the integrity and strength of a pressure vessel by subjecting it to a high pressure water test. The hydrostatic test is performed by filling the vessel with water and applying a pressure that is higher than the design pressure of the vessel. The pressure is held for a specified period of time and the vessel is carefully inspected for leaks, deformations, or other signs of damage. The hydrostatic test is typically performed after the vessel has been fabricated and before it is put into service to ensure it is safe and fit for its intended purpose.

Question: What is a pressure vessel design calculation?

Answer: A pressure vessel design calculation is a series of mathematical equations that are used to determine the size, shape, and thickness of a pressure vessel based on the design pressure and temperature, the materials of construction, and other factors. The design calculation is an important step in the design process and is used to ensure that the vessel is safe and fit for its intended purpose. The design calculation may include calculations for the vessel shell, heads, nozzles, joints, supports, and other components. The design calculation may be performed using hand calculations or computer software, such as a finite element analysis (FEA) program.

Question: What is a pressure vessel fabrication?

Answer: Pressure vessel fabrication is the process of constructing a pressure vessel from raw materials using a series of manufacturing techniques, such as cutting, forming, welding, and assembly. The fabrication process involves the preparation of the materials, the layout and marking of the parts, the cutting and forming of the parts, and the welding and assembly of the parts. The fabrication process is typically carried out in a fabrication shop by skilled craftsmen using specialized equipment, such as lathes, mills, and welding machines. The fabrication process must be carefully controlled to ensure that the vessel is of high quality and meets the required design and safety standards.

Question: What is a pressure vessel inspection report?

Answer: A pressure vessel inspection report is a document that summarizes the findings of a pressure vessel inspection and provides recommendations for any repairs or replacements that may be needed. The inspection report typically includes a description of the vessel, the inspection method and techniques used, the results of the inspection, and any defects or deficiencies that were found. The inspection report may also include a list of recommended repairs or replacements, an estimate of the cost of the repairs, and a timeline for completing the repairs. The inspection report is an important tool for managing the maintenance and repair of pressure vessels and ensuring that they are safe and fit for service.

Question: What is a pressure vessel visual inspection?

Answer: A pressure vessel visual inspection is a type of inspection that involves examining the exterior and interior of a pressure vessel using the naked eye or visual aids, such as magnifying glasses or boroscopes. The visual inspection is used to identify any visible defects or abnormalities in the vessel, such as cracks, dents, corrosion, or deformations. The visual inspection is usually the first step in a pressure vessel inspection and is typically followed by more detailed inspection methods, such as non-destructive testing or destructive testing.

Question: What is a pressure vessel non-destructive testing?

Answer: Pressure vessel non-destructive testing (NDT) is a type of inspection that involves evaluating the condition of a pressure vessel without causing any damage or alterations to the vessel. NDT methods include techniques such as ultrasonic testing, magnetic particle testing, radiographic testing, and eddy current testing, which are used to detect internal or hidden defects in the vessel. NDT is typically used to supplement visual inspection and to provide a more complete evaluation of the vessel's condition. NDT is often used to inspect pressure vessels that are in-service or have limited accessibility.

Question: What is a pressure vessel destructive testing?

Answer: Pressure vessel destructive testing (DT) is a type of inspection that involves evaluating the condition of a pressure vessel by subjecting it to extreme loads or conditions that may cause damage or failure. DT methods include techniques such as tensile testing, impact testing, fatigue testing, and hardness testing, which are used to evaluate the strength, toughness, and ductility of the vessel materials. DT is typically used to supplement non-destructive testing and to provide a more comprehensive evaluation of the vessel's condition. DT is often used to verify the accuracy of design calculations and to validate the materials of construction.

Question: What is a pressure vessel repair?

Answer: A pressure vessel repair is a process of restoring a damaged or defective pressure vessel to a safe and functional condition. The repair may involve replacing damaged parts, such as nozzles or flanges, welding or grinding out defects, or applying protective coatings or linings. The repair must be performed by qualified personnel using appropriate repair methods and materials to ensure the vessel is safe and fit for service. The repair may be required due to corrosion, fatigue, overpressure, or other causes of damage.

Question: What is a pressure vessel refurbishment?

Answer: Pressure vessel refurbishment is a process of rejuvenating an old or obsolete pressure vessel to extend its operational life or improve its performance. The refurbishment may involve repairs, upgrades, or modifications to the vessel, such as replacing outdated components, adding new features, or improving the efficiency or reliability of the vessel. The refurbishment may be performed to meet changing operational needs, to extend the design life of the vessel, or to improve the safety or performance of the vessel. The refurbishment must be carefully planned and executed to ensure the vessel is safe and fit for service.

Question: What is a pressure vessel inspection interval?

Answer: A pressure vessel inspection interval is the recommended frequency at which a pressure vessel should be inspected to ensure it is safe and fit for service. The inspection interval is typically based on the age, condition, and type of the vessel, as well as the operating environment and the level of risk associated with the vessel. The inspection interval may be expressed in terms of the number of operating hours, the number of operating cycles, or the number of years of service. The inspection interval is typically specified in the vessel design code or other industry standards.

Question: What is a pressure vessel classification society?

Answer: A pressure vessel classification society is an organization that provides certification and inspection services for pressure vessels and other types of pressure equipment. The classification society sets standards and guidelines for the design, construction, and operation of pressure vessels and helps ensure that they are safe and fit for service. Some common pressure vessel classification societies include the American Bureau of Shipping (ABS), the Lloyds Register (LR), and the Det Norske Veritas (DNV). The classification society may also provide technical support and consulting services to vessel owners and operators.

Question: What is a pressure vessel certification?

Answer: Pressure vessel certification is the process of verifying that a pressure vessel meets the requirements of a specific design code or standard and is fit for service. The certification process typically involves reviewing the design calculations, inspecting the vessel during fabrication, and testing the vessel to ensure it meets the required specifications. The certification may be provided by a classification society, a government agency, or another accredited organization. The certification is important because it helps ensure the safety and reliability of the vessel and its contents.

Question: What is a pressure vessel inspection agency?

Answer: A pressure vessel inspection agency is an organization that provides inspection and certification services for pressure vessels and other types of pressure equipment. The inspection agency is responsible for evaluating the condition of the vessel and ensuring that it meets the required design and safety standards. The inspection agency may be a classification society, a government agency, or another accredited organization. The inspection agency may provide inspection services at various stages of the vessel's life cycle, including during fabrication, installation, operation, and decommissioning.

Question: What is a pressure vessel installation?

Answer: Pressure vessel installation is the process of installing a pressure vessel in its intended location and preparing it for operation. The installation process typically involves securing the vessel to its supports, attaching piping and other equipment, and performing any necessary testing or commissioning. The installation process must be carefully planned and executed to ensure the safety and reliability of the vessel. The installation may be performed by the manufacturer or a contractor who is qualified to install pressure vessels.

Question: What is a pressure vessel decommissioning?

Answer: Pressure vessel decommissioning is the process of retiring a pressure vessel from service and preparing it for disposal or reuse. The decommissioning process may involve draining and cleaning the vessel, removing any hazardous materials, and preparing the vessel for transportation. The decommissioning process must be carefully planned and executed to ensure the safety of the vessel and its contents and to minimize the impact on the environment. The decommissioning process may be performed by the owner or operator of the vessel or by a contractor who is qualified to decommission pressure vessels.

Question: What is a pressure vessel maintenance?

Answer: Pressure vessel maintenance is the process of keeping a pressure vessel in good working order and ensuring that it is safe and fit for service. The maintenance process may involve inspections, repairs, refurbishments, or upgrades to the vessel. The maintenance process is typically performed on a regular basis, such as annually or based on the inspection interval, to ensure the vessel is operating correctly and to prevent problems from occurring. The maintenance process may be performed by the owner or operator of the vessel or by a contractor who is qualified to maintain pressure vessels.

Question: What is a pressure vessel design specification?

Answer: A pressure vessel design specification is a document that outlines the requirements for the design, construction, and testing of a pressure vessel. The design specification typically includes information about the vessel's intended use, the design pressure and temperature, the materials of construction, the structural requirements, the fabrication and testing requirements, and the inspection and certification requirements. The design specification is an important tool for ensuring that the vessel is safe and fit for its intended purpose and meets the required design and safety standards. The design specification may be based on a design code or other industry standards.

Question: What is a pressure vessel design drawing?

Answer: A pressure vessel design drawing is a technical drawing that shows the dimensions, shape, and details of a pressure vessel. The design drawing may include views of the vessel from different angles, sections through the vessel, and details of the components, such as nozzles, flanges, and supports. The design drawing is an important tool for communicating the design of the vessel to the fabricator and for ensuring that the vessel is built according to the design specifications. The design drawing may be prepared using computer-aided design (CAD) software or by hand using drafting instruments.

Question: What is a pressure vessel safety valve?

Answer: A pressure vessel safety valve is a device that is installed on a pressure vessel to protect it from overpressure. The safety valve is designed to open at a predetermined pressure and release excess pressure to prevent the vessel from exceeding its design pressure. The safety valve is typically a spring-loaded valve that is actuated by the pressure in the vessel. The safety valve must be carefully selected and sized to ensure it can handle the maximum expected flow rate and relieving capacity. The safety valve must also be properly installed and maintained to ensure it functions properly in the event of an overpressure situation.

Question: What is a pressure vessel inspection checklist?

Answer: A pressure vessel inspection checklist is a list of items that are inspected during a pressure vessel inspection. The checklist may include items such as the vessel identification, the materials of construction, the design pressure and temperature, the corrosion protection, the nozzles and flanges, the supports and attachments, and the safety devices. The inspection checklist is used to ensure that all relevant aspects of the vessel are inspected and to document the results of the inspection. The inspection checklist may be based on a design code or industry standard and may be used as part of a formal inspection report.

Question: What is a pressure vessel support?

Answer: A pressure vessel support is a device that is used to hold a pressure vessel in place and transfer the load of the vessel to the foundation or the surrounding structure. The support may be a structural element, such as a pedestal or a beam, or a spring or other type of load-bearing device. The support must be able to withstand the weight of the vessel and any additional loads, such as wind or seismic loads. The support must also be compatible with the materials of construction of the vessel and the operating temperature of the vessel.

Question: What are some common materials used for pressure vessel fabrication?

Answer: Common materials used for pressure vessel fabrication include carbon steel, stainless steel, aluminum, and various alloys.

The choice of material depends on the specific application and operating conditions of the pressure vessel. Factors to consider when selecting a material include the pressure and temperature of the vessel's contents, the corrosiveness of the environment, and the strength and ductility requirements of the vessel.

Question: What are the key considerations for designing a pressure vessel?

Answer: Some key considerations for designing a pressure vessel include determining the required pressure and temperature ratings, selecting an appropriate material, ensuring structural integrity and stability, and ensuring that the vessel complies with relevant codes and standards. Other factors to consider may include the size and shape of the vessel, the type of closure and opening, and the presence of any internal components or attachments.

Question: What are some common fabrication techniques used for pressure vessels?

Answer: Some common fabrication techniques used for pressure vessels include rolling, forging, stamping, and welding. The choice of technique will depend on the specific material and design of the vessel, as well as the equipment and capabilities of the fabricator. For example, rolling may be used to create cylindrical shapes from sheet metal, while forging may be used to create complex shapes from solid bars of metal.

Welding is a common method for joining the various parts of a pressure vessel, and can be done using various techniques such as TIG, MIG, and stick welding.

Question: How are pressure vessels tested and inspected during fabrication?

Answer: There are several methods used to test and inspect pressure vessels during fabrication to ensure that they are safe and meet the required specifications. These methods may include visual inspection, dimensional inspection, pressure testing, and nondestructive testing (NDT).

Visual inspection involves examining the surface and welds of the vessel for defects or deviations from the design. Dimensional inspection involves measuring the dimensions of the vessel and its components to ensure that they meet the specified tolerances. Pressure testing involves filling the vessel with a pressurized fluid and measuring the pressure to ensure that it meets the required pressure rating. Nondestructive testing (NDT) involves using techniques such as radiography, ultrasonics, and magnetic particle inspection to examine the internal and external surfaces of the vessel for defects without damaging the vessel.

Question: What are some common applications for pressure vessels?

Answer: Pressure vessels are used in a wide range of industries and applications. Some common examples include storage tanks for gases and liquids, boilers and steam generators, compressed air tanks, chemical reactors, and separators. Pressure vessels are also used in many other industries such as food and beverage processing, pharmaceuticals, and petroleum refining.

CHAPTER FIVE

Questions and answers related to Pressure vessel formulas

What is the formula for calculating the minimum thickness of a cylindrical pressure vessel?

The minimum thickness of a cylindrical pressure vessel can be calculated using the following formula:

t = PD/2(SE + PY)

where:

- t = minimum thickness
- P = internal pressure
- D = diameter of the vessel
- SE = allowable stress in the material of the vessel
- Y = a factor that accounts for the shape and loading of the vessel

What is the formula for calculating the maximum allowable working pressure of a pressure vessel with an elliptical head?

The maximum allowable working pressure of a pressure vessel with an elliptical head can be calculated using the following formula:

MAWP = 2SE/E + PY

where:

- MAWP = maximum allowable working pressure
- SE = allowable stress in the material of the vessel
- E = a factor that accounts for the shape of the head
- Y = a factor that accounts for the loading of the vessel

How do I calculate the volume of a pressure vessel in the shape of a sphere?

The volume of a spherical pressure vessel can be calculated using the following formula:

$V = 4/3\pi r^3$

where:

- V = volume
- r = radius of the sphere

How do I calculate the surface area of a pressure vessel in the shape of a cylinder?

The surface area of a cylindrical pressure vessel can be calculated using the following formula:

$A = 2\pi r(r + l)$

where:

- A = surface area
- r = radius of the cylinder
- l = length of the cylinder

What is the formula for calculating the stress on a pressure vessel subjected to an internal pressure?

The stress on a pressure vessel subjected to an internal pressure can be calculated using the following formula:

$\sigma = P(r/t)$

where:

- σ = stress on the vessel
- P = internal pressure

- r = radius of the vessel
- t = thickness of the vessel

What is the formula for calculating the required thickness of a conical pressure vessel?

The required thickness of a conical pressure vessel can be calculated using the following formula:

t = PD/2(SE + PY)

where:

- t = required thickness
- P = internal pressure
- D = diameter of the vessel at the largest end
- SE = allowable stress in the material of the vessel
- Y = a factor that accounts for the shape and loading of the vessel

How do I calculate the volume of a pressure vessel in the shape of a frustum of a cone?

The volume of a frustum of a cone-shaped pressure vessel can be calculated using the following formula:

V = 1/3πh(R^2 + r^2 + Rr)

where:

- V = volume
- h = height of the frustum
- R = radius of the larger base
- r = radius of the smaller base

How do I calculate the surface area of a pressure vessel in the shape of a cone?

The surface area of a cone-shaped pressure vessel can be calculated using the following formula:

A = πr(l + r)

where:

- A = surface area
- r = radius of the base of the cone
- l = slant height of the cone

What is the formula for calculating the maximum allowable working pressure of a pressure vessel with a hemispherical head?

The maximum allowable working pressure of a pressure vessel with a hemispherical head can be calculated using the following formula:

MAWP = 2SE/E + PY

where:

- MAWP = maximum allowable working pressure
- SE = allowable stress in the material of the vessel
- E = a factor that accounts for the shape of the head
- Y = a factor that accounts for the loading of the vessel

How do I calculate the volume of a pressure vessel in the shape of a cylinder with both ends closed (a "bullet" shape)?

The volume of a "bullet"-shaped pressure vessel can be calculated using the following formula:

V = πr^2l

where:

- V = volume
- r = radius of the cylinder
- l = length of the cylinder

How do I calculate the surface area of a pressure vessel in the shape of a cylinder with both ends closed?

The surface area of a "bullet"-shaped pressure vessel can be calculated using the following formula:

$A = 2\pi r(r + l)$

where:

- A = surface area
- r = radius of the cylinder
- l = length of the cylinder

What is the formula for calculating the maximum allowable working pressure of a pressure vessel with a torispherical head?

The maximum allowable working pressure of a pressure vessel with a torispherical head can be calculated using the following formula:

MAWP = 2SE/E + PY

where:

- MAWP = maximum allowable working pressure
- SE = allowable stress in the material of the vessel
- E = a factor that accounts for the shape of the head
- Y = a factor that accounts for the loading of the vessel

How do I calculate the volume of a pressure vessel in the shape of a pyramid with a triangular base?

The volume of a pyramid-shaped pressure vessel with a triangular base can be calculated using the following formula:

V = (Bh)/3

where:

- V = volume
- B = area of the base of the pyramid
- h = height of the pyramid

How do I calculate the surface area of a pressure vessel in the shape of a pyramid with a triangular base?

The surface area of a pyramid-shaped pressure vessel with a triangular base can be calculated using the following formula:

A = B + (3l^2)/s

where:

- A = surface area
- B = area of the base of the pyramid
- l = slant height of the pyramid
- s = side length of the base of the pyramid

What is the formula for calculating the maximum allowable working pressure of a pressure vessel with a dished head?

The maximum allowable working pressure of a pressure vessel with a dished head can be calculated using the following formula:

MAWP = 2SE/E + PY

where:

- MAWP = maximum allowable working pressure
- SE = allowable stress in the material of the vessel
- E = a factor that accounts for the shape of the head
- Y = a factor that accounts for the loading of the vessel

How do I calculate the volume of a pressure vessel in the shape of a cylinder with one closed end and one open end (a "truncated cone" shape)?

The volume of a "truncated cone"-shaped pressure vessel can be calculated using the following formula:

V = π/3(R^2 + r^2 + Rr)h

where:

- V = volume
- R = radius of the larger base (the closed end)
- r = radius of the smaller base (the open end)
- h = height of the truncated cone

How do I calculate the surface area of a pressure vessel in the shape of a cylinder with one closed end and one open end?

The surface area of a "truncated cone"-shaped pressure vessel can be calculated using the following formula:

A = π(R + r)(R + r + l)

where:

- A = surface area
- R = radius of the larger base (the closed end)
- r = radius of the smaller base (the open end)
- l = slant height of the truncated cone

What is the formula for calculating the maximum allowable working pressure of a pressure vessel with a flanged and dished head?

The maximum allowable working pressure of a pressure vessel with a flanged and dished head can be calculated using the following formula:

MAWP = 2SE/E + PY

where:

- MAWP = maximum allowable working pressure
- SE = allowable stress in the material of the vessel
- E = a factor that accounts for the shape of the head
- Y = a factor that accounts for the loading of the vessel

How do I calculate the volume of a pressure vessel in the shape of a cylinder with one end closed and one end open, and a cylindrical neck extending from the open end?

The volume of a pressure vessel with a cylindrical neck can be calculated by treating the vessel as two separate shapes: a cylinder with a closed end and a cylinder with an open end. The volume of the closed-end cylinder can be calculated using the formula for a "bullet"-shaped pressure vessel:

V1 = πr^2l

The volume of the open-end cylinder (the neck) can be calculated using the formula for a cylinder with one open end:

V2 = πr^2h

The total volume of the vessel is the sum of the volumes of the two cylinders:

$V = V1 + V2$

where:

- V = volume
- r = radius of the cylinder
- l = length of the closed-end cylinder
- h = height of the neck

How do I calculate the surface area of a pressure vessel in the shape of a cylinder with one end closed and one end open, and a cylindrical neck extending from the open end?

The surface area of a pressure vessel with a cylindrical neck can be calculated by treating the vessel as two separate shapes: a cylinder with a closed end and a cylinder with an open end. The surface area of the closed-end cylinder can be calculated using the formula for a "bullet"-shaped pressure vessel:

$A1 = 2\pi r(r + l)$

The surface area of the open-end cylinder (the neck) can be calculated using the formula for a cylinder with one open end:

$A2 = 2\pi r^2 + 2\pi rh$

The total surface area of the vessel is the sum of the surface areas of the two cylinders:

$A = A1 + A2$

where:

- A = surface area
- r = radius of the cylinder
- l = length of the closed-end cylinder
- h = height of the neck

What is the formula for calculating the wall thickness of a pressure vessel?

The formula for calculating the wall thickness of a pressure vessel is: t = PS / (2S*y + DP), where P is the internal design pressure, S is the allowable stress, y is the joint efficiency factor, and

DP is the design pressure.

How is the maximum allowable pressure for a pressure vessel calculated?

The maximum allowable pressure for a pressure vessel is calculated using the following formula: $P = 2Sy / (t/r - 1/r)$, where P is the maximum allowable pressure, S is the allowable stress, y is the joint efficiency factor, t is the wall thickness, and r is the radius of the vessel.

What is the formula for calculating the volume of a pressure vessel?

The formula for calculating the volume of a pressure vessel is: $V = \pi r^2 L$, where V is the volume, r is the radius of the vessel, and L is the length of the vessel.

How is the weight of a pressure vessel calculated?

The weight of a pressure vessel can be calculated using the following formula: $W = V\rho g$, where W is the weight, V is the volume of the vessel, ρ is the density of the material from which the vessel is made, and g is the acceleration due to gravity.

What is the formula for calculating the required thickness of a cylindrical pressure vessel subjected to an internal pressure?

The required thickness of a cylindrical pressure vessel subjected to an internal pressure can be calculated using the following formula: $t = PD / (2*SE - 0.6P)$, where t is the required thickness, P is the internal design pressure, D is the diameter of the vessel, and SE is the allowable stress in the material of the vessel.

How is the maximum allowable working pressure of a pressure vessel calculated using the Barlow's formula?

The maximum allowable working pressure of a pressure vessel can be calculated using Barlow's formula: $P = 2Sy / (t/D + 1/2*y)$, where P is the maximum allowable working pressure, S is the allowable stress in the material of the vessel, y is the joint efficiency factor, t is the wall thickness, and D is the diameter of the vessel.

What is the formula for calculating the required thickness of a spherical pressure vessel subjected to an internal pressure?

The required thickness of a spherical pressure vessel subjected to an internal pressure can be calculated using the following formula: t = PD / (2*SE - 0.2P), where t is the required thickness, P is the internal design pressure, D is the diameter of the sphere, and SE is the allowable stress in the material of the vessel.

How is the hoop stress in a cylindrical pressure vessel calculated?

The hoop stress in a cylindrical pressure vessel can be calculated using the following formula: σ_h = P/t, where σ_h is the hoop stress, P is the internal design pressure, and t is the wall thickness of the vessel.

What is the formula for calculating the required thickness of a conical pressure vessel subjected to an internal pressure?

The required thickness of a conical pressure vessel subjected to an internal pressure can be calculated using the following formula: t = PD / (2SE - 0.2Pcos(α)), where t is the required thickness, P is the internal design pressure, D is the diameter of the base of the cone, SE is the allowable stress in the material of the vessel, and α is the half angle of the cone.

How is the maximum allowable working pressure of a pressure vessel with an elliptical head calculated?

The maximum allowable working pressure of a pressure vessel with an elliptical head can be calculated using the following formula: P = 2SEy / (t/r + 1/r), where P is the maximum allowable working pressure, SE is the allowable stress in the material of the vessel, y is the joint efficiency factor, t is the wall thickness, and r is the radius of the curvature of the head.

What is the formula for calculating the required thickness of a hemispherical pressure vessel subjected to an internal pressure?

The required thickness of a hemispherical pressure vessel subjected to an internal pressure can be calculated using the following formula: t = PD / (2*SE - 0.2P), where t is the required thickness, P is the internal design pressure, D is the diameter of the hemisphere, and SE is the allowable stress in the material of the vessel.

How is the maximum allowable working pressure of a pressure vessel with a torispherical head calculated?

The maximum allowable working pressure of a pressure vessel with a torispherical head can be calculated using the following formula: $P = 2SEy / (t/r_d + 1/r_f - 1/r_d)$, where P is the maximum allowable working pressure, SE is the allowable stress in the material of the vessel, y is the joint efficiency factor, t is the wall thickness, r_d is the dish radius, and r_f is the knuckle radius.

What is the formula for calculating the required thickness of a pressure vessel with an hemispherical head subjected to an internal pressure?

The required thickness of a pressure vessel with an hemispherical head subjected to an internal pressure can be calculated using the following formula: $t = PD / (2*SE - 0.2P)$, where t is the required thickness, P is the internal design pressure, D is the diameter of the hemisphere, and SE is the allowable stress in the material of the vessel.

How is the maximum allowable working pressure of a pressure vessel with a flanged and dished head calculated?

The maximum allowable working pressure of a pressure vessel with a flanged and dished head can be calculated using the following formula: $P = 2SEy / (t/r_d + 1/r_f - 1/r_d)$, where P is the maximum allowable working pressure, SE is the allowable stress in the material of the vessel, y is the joint efficiency factor, t is the wall thickness, r_d is the dish radius, and r_f is the flange radius.

What is the formula for calculating the required thickness of a pressure vessel with a flanged and dished head subjected to an internal pressure?

The required thickness of a pressure vessel with a flanged and dished head subjected to an internal pressure can be calculated using the following formula: $t = PD / (2*SE - 0.2P)$, where t is the required thickness, P is the internal design pressure, D is the diameter of the dish, and SE is the allowable stress in the material of the vessel.

How is the maximum allowable working pressure of a pressure vessel with a torispherical head calculated using the ASME code?

The maximum allowable working pressure of a pressure vessel with a torispherical head calculated using the ASME code can be calculated using the following formula: $P = 2SEy / (t/r_d + 1/r_f - 0.17/r_d)$, where P is the maximum allowable working pressure, SE is the allowable stress in the material of the vessel, y is the joint efficiency factor, t is the wall thickness, r_d is the dish radius, and r_f is the knuckle radius.

$y / (t/r + \cos(\alpha)/r)$, where P is the maximum allowable working pressure, SE is the allowable stress in the material of the vessel, y is the joint efficiency factor, t is the wall thickness, r is the radius of the base of the cone, and α is the half angle of the cone.

CHAPTER SIX

Conclusion:

In conclusion, this book on pressure vessel questions and answers has provided a comprehensive overview of the key concepts and issues related to the design, construction, operation, and maintenance of pressure vessels.

We have covered a range of topics, including the different types of pressure vessels, the materials and construction methods used to build them, the various design and safety considerations that must be taken into account, and the various testing and inspection procedures that are used to ensure their safe operation.

We hope that this book has provided a valuable resource for anyone interested in learning more about pressure vessels and their role in various industries. Whether you are a student, engineer, or professional working in the field, we believe that the information and insights contained in these pages will help you to better understand and appreciate the importance of pressure vessels and the critical role they play in a wide range of applications.

Thank you for reading, and we hope that you have found this book to be useful and informative.

Disclaimer:

Disclaimer:

The information contained in "Pressure Vessel interview Questions and Answers" is for general information purposes only. The information is provided by [Chetan Singh] and while we endeavour to keep the information up to date and correct, we make no representations or warranties of any kind, express or implied, about the completeness, accuracy, reliability, suitability or availability with respect to the book or the information, products, services, or related graphics contained in the book for any purpose. Any reliance you place on such information is therefore strictly at your own risk.

In no event will we be liable for any loss or damage including without limitation, indirect or consequential loss or damage, or any loss or damage whatsoever arising from loss of data or profits arising out of, or in connection with, the use of this book.

Book References: Wikipedia, ASME, API, Google and relevant websites.

Note: please read updated code, standards, rules and regulations from relevant organizations like: ASME BPVC, API, PED, and AS 4041, etc.,

Thank You

Printed by Libri Plureos GmbH in Hamburg,
Germany